# PET–CT FOR THE MANAGEMENT
# OF CANCER PATIENTS

The following States are Members of the International Atomic Energy Agency:

AFGHANISTAN
ALBANIA
ALGERIA
ANGOLA
ANTIGUA AND BARBUDA
ARGENTINA
ARMENIA
AUSTRALIA
AUSTRIA
AZERBAIJAN
BAHAMAS
BAHRAIN
BANGLADESH
BARBADOS
BELARUS
BELGIUM
BELIZE
BENIN
BOLIVIA, PLURINATIONAL
  STATE OF
BOSNIA AND HERZEGOVINA
BOTSWANA
BRAZIL
BRUNEI DARUSSALAM
BULGARIA
BURKINA FASO
BURUNDI
CAMBODIA
CAMEROON
CANADA
CENTRAL AFRICAN
  REPUBLIC
CHAD
CHILE
CHINA
COLOMBIA
COMOROS
CONGO
COSTA RICA
CÔTE D'IVOIRE
CROATIA
CUBA
CYPRUS
CZECH REPUBLIC
DEMOCRATIC REPUBLIC
  OF THE CONGO
DENMARK
DJIBOUTI
DOMINICA
DOMINICAN REPUBLIC
ECUADOR
EGYPT
EL SALVADOR
ERITREA
ESTONIA
ESWATINI
ETHIOPIA
FIJI
FINLAND
FRANCE
GABON
GEORGIA

GERMANY
GHANA
GREECE
GRENADA
GUATEMALA
GUYANA
HAITI
HOLY SEE
HONDURAS
HUNGARY
ICELAND
INDIA
INDONESIA
IRAN, ISLAMIC REPUBLIC OF
IRAQ
IRELAND
ISRAEL
ITALY
JAMAICA
JAPAN
JORDAN
KAZAKHSTAN
KENYA
KOREA, REPUBLIC OF
KUWAIT
KYRGYZSTAN
LAO PEOPLE'S DEMOCRATIC
  REPUBLIC
LATVIA
LEBANON
LESOTHO
LIBERIA
LIBYA
LIECHTENSTEIN
LITHUANIA
LUXEMBOURG
MADAGASCAR
MALAWI
MALAYSIA
MALI
MALTA
MARSHALL ISLANDS
MAURITANIA
MAURITIUS
MEXICO
MONACO
MONGOLIA
MONTENEGRO
MOROCCO
MOZAMBIQUE
MYANMAR
NAMIBIA
NEPAL
NETHERLANDS
NEW ZEALAND
NICARAGUA
NIGER
NIGERIA
NORTH MACEDONIA
NORWAY
OMAN
PAKISTAN

PALAU
PANAMA
PAPUA NEW GUINEA
PARAGUAY
PERU
PHILIPPINES
POLAND
PORTUGAL
QATAR
REPUBLIC OF MOLDOVA
ROMANIA
RUSSIAN FEDERATION
RWANDA
SAINT KITTS AND NEVIS
SAINT LUCIA
SAINT VINCENT AND
  THE GRENADINES
SAMOA
SAN MARINO
SAUDI ARABIA
SENEGAL
SERBIA
SEYCHELLES
SIERRA LEONE
SINGAPORE
SLOVAKIA
SLOVENIA
SOUTH AFRICA
SPAIN
SRI LANKA
SUDAN
SWEDEN
SWITZERLAND
SYRIAN ARAB REPUBLIC
TAJIKISTAN
THAILAND
TOGO
TONGA
TRINIDAD AND TOBAGO
TUNISIA
TÜRKİYE
TURKMENISTAN
UGANDA
UKRAINE
UNITED ARAB EMIRATES
UNITED KINGDOM OF
  GREAT BRITAIN AND
  NORTHERN IRELAND
UNITED REPUBLIC
  OF TANZANIA
UNITED STATES OF AMERICA
URUGUAY
UZBEKISTAN
VANUATU
VENEZUELA, BOLIVARIAN
  REPUBLIC OF
VIET NAM
YEMEN
ZAMBIA
ZIMBABWE

The Agency's Statute was approved on 23 October 1956 by the Conference on the Statute of the IAEA held at United Nations Headquarters, New York; it entered into force on 29 July 1957. The Headquarters of the Agency are situated in Vienna. Its principal objective is "to accelerate and enlarge the contribution of atomic energy to peace, health and prosperity throughout the world".

IAEA HUMAN HEALTH SERIES No. 45

# PET–CT FOR THE MANAGEMENT OF CANCER PATIENTS

## A REVIEW OF THE EXISTING EVIDENCE

INTERNATIONAL ATOMIC ENERGY AGENCY
VIENNA, 2023

# COPYRIGHT NOTICE

© IAEA, 2023

Printed by the IAEA in Austria
January 2023
STI/PUB/1993

**IAEA Library Cataloguing in Publication Data**

Names: International Atomic Energy Agency.
Title: PET–CT for the management of cancer patients : a review of the existing evidence / International Atomic Energy Agency.
Description: Vienna : International Atomic Energy Agency, 2023. | Series: IAEA human health series, ISSN 2075–3772 ; no. No. 45 | Includes bibliographical references.
Identifiers: IAEAL 22-01495 | ISBN 978–92–0–118622–5 (paperback : alk. paper) | ISBN 978–92–0–118422–1 (pdf) | ISBN 978–92–0–118522–8 (epub)
Subjects: LCSH: Tomography, Emission. | Cancer — Patients. | Nuclear medicine.
Classification: UDC 616-073 | STI/PUB/1993

# FOREWORD

The global incidence of cancer is increasing in both developed and developing countries and will become an increasing health burden in the coming decades. This rise in the cancer rate will bring with it challenges for health care systems, clinicians, and patients and their families. Technologies that improve the decision making process and optimize treatment have the potential to benefit society as a whole.

The purpose of this publication, predominantly aimed at policy makers, is to develop a consensus — based on the existing evidence — on the value and the main indications of hybrid imaging using positron emission tomography (PET) combined with computed tomography (CT) in the management of patients affected by cancer. Indeed, PET–CT is considered to be a growing part of the health care landscape due to the rising prevalence of non-communicable diseases, the need for early and accurate diagnostic methods, the technological developments in both hardware and software, the availability of new tracers and its acceptance in emerging markets. Fluorodeoxyglucose (FDG) PET–CT has earned global recognition as a significant tool in the modern management of cancer patients. However, FDG has limitations in its ability to assess several prevalent tumours, such as those produced by prostate cancer. In addition, new therapeutic options available today in the management of cancer have underscored the need to assess tumour characteristics other than metabolism. Therefore, there has been a pressing need for the development and clinical assessment of additional PET radiopharmaceuticals that can enable the imaging and precise characterization of various aspects of a wide range of malignant tumours.

While the use of PET–CT is a standard of care in oncological practice in many developed countries, it is still limited in many low to middle income nations. Based on these considerations, the IAEA recognizes the need to make reliable information widely available to support Member States in the use of PET–CT.

To achieve this goal, the IAEA convened an expert consultant group to review the previous publication, Appropriate Use of FDG PET for the Management of Cancer Patients, IAEA Human Health Series No. 9, in view of the most recent evidence. In the present publication, we focus our attention on highlighting the main indications of FDG and non-FDG radiopharmaceuticals in the management of cancer patients, based on the current clinical evidence.

The recommendations included here to promote the optimal use of PET–CT imaging procedures in oncology considered the most recent developments of PET radiopharmaceuticals. These broad recommendations cannot be applied rigidly to all patients in all clinical settings, but might be considered as a valid basis for tumour board discussions.

The IAEA officer responsible for this publication was F. Giammarile of the Division of Human Health.

# CONTENTS

# 1. INTRODUCTION

## 1.1. BACKGROUND

A multidisciplinary approach to cancer management is increasingly required to provide personalized evidence based patient care. Appropriateness is a guiding principle to justify any new health care intervention: from the use of new drugs or new treatment modalities to the implementation of new diagnostic procedures. The concept of appropriateness, with a decision aid for its assessment, provides clinicians and policy makers with a tool to determine which diagnostic investigations and therapies ought to be implemented.

Appropriateness of diagnostic investigations has been defined in terms of clinical utility and may also be used to assist in the allocation of limited resources in health care. There is, however, a definite risk that new interventions will be underutilized because they are viewed by policy makers as inappropriate or inappropriately expensive. This could be due to a narrow interpretation of appropriateness that is based solely on the cost of the intervention, isolated from the potential cost savings derived from its use. Therefore, there might be a series of interventions, services and health services of proven effectiveness that are widely underutilized, whose necessary implementation requires, at least in the short and medium terms, an increase in costs.

Policy makers need to accept that the main aim of appropriateness is not cost reduction, but rather optimization of health resource allocation, recognizing the unfavourable consequences of failure to implement innovations of proven effectiveness. It is only through acceptance of this perspective that innovations of proven effectiveness will be introduced for the benefit of both individuals and society.

In nuclear medicine, positron emission tomography (PET) technology has emerged as an integral part of patient management in oncology and is now the standard of care for most tumour types. While not a new concept, in recent years theranostics, the combination of diagnostic and therapy using similar molecules, has enjoyed a revival. This is largely owing to the integration of new PET tracers with radionuclide agents for therapy [1–4].

## 1.2. OBJECTIVE

The purpose of this publication is to revise and update previous guidelines through consensus based on current evidence, in order to make health care providers and policy makers aware of the value and the recommended use of

1

positron emission tomography combined with computed tomography (PET–CT) in the management of cancer patients [5]. Guidance provided here, describing good practices, represents expert opinion but does not constitute recommendations made on the basis of a consensus of Member States.

## 1.3. SCOPE

This publication provides a good overview of all the common indications of PET–CT in clinical oncology. It is hoped that this publication will be beneficial to medical professionals from IAEA Member States for learning and teaching purposes.

## 1.4. STRUCTURE

Indications for the use of PET–CT in the management of multiple cancers are outlined in Section 2 and presented in more detail in subsequent sections. Several possible indications are considered for each type of cancer, with recommendations given for each indication in Section 3.

# 2. CLINICAL SCENARIOS FOR PET–CT INDICATIONS

## 2.1. SEARCH STRATEGY

A search of the available scientific publications was initially confined to systematic reviews on the use of PET–CT in oncology published prior to 2009 for Human Health Series No. 9 [5]. This version will incorporate additional literature relevant to recent improvements in the availability of newer PET radiopharmaceuticals and updated indications for 2-fluoro-2-deoxy-D-glucose (FDG) PET [4].

## 2.2. LIST OF INCLUDED RADIOPHARMACEUTICALS

### 2.2.1. Acetate: [$^{11}$C]acetate

Acetate is integrated into the intracellular phosphatidylcholine membrane microdomains (dominant pathway in cancer cells).

### 2.2.2. Choline: [$^{11}$C]choline chloride, [$^{18}$F]fluoroethylcholine or [$^{18}$F]fluoromethylcholine

Choline is an important component of phospholipids in cell membranes. Tissues with increased cell turnover frequently exhibit increased uptake of choline.

### 2.2.3. Somatostatin analogues such as [$^{68}$Ga]DOTA-NOC, -TOC or -TATE

Synthetic somatostatin peptides with a long biological half-life display high specific affinity for somatostatin receptors expressed on the cellular surface of neuroendocrine tumours.

### 2.2.4. FDOPA: 6-[$^{18}$F]fluoro-L-DOPA

The fate of radiolabelled fluoro-l-dihydroxyphenylalanine (FDOPA) mirrors all stages of native dihydroxyphenylalanine (DOPA) transport, storage and metabolism.

### 2.2.5. FDG: [$^{18}$F]fluorodeoxyglucose

Uptake of the glucose analogue FDG is closely correlated with cancer metabolism and proliferation rate.

### 2.2.6. FES: [$^{18}$F]fluoroestradiol

FES binds to the oestrogen receptors on the tumour cell surface as well as intratumoural receptors in oestrogen receptor positive tumours

### 2.2.7. FET: [$^{18}$F]fluoroethyltyrosine

FET is a radiolabelled amino acid but is not incorporated into proteins and remains trapped in cells after uptake. Certain tumour subtypes exhibit high uptake of FET.

### 2.2.8. Fluoride: [$^{18}$F]sodium fluoride

The intensity of deposition of fluoride ions in the bone matrix reflects bone remodelling and blood flow. Note that fluoride PET is not covered in this guideline because its use is universally relevant to all bone seeking tumours.

### 2.2.9. Methionine: [$^{11}$C]methionine

Methionine is an essential amino acid involved in angiogenesis and demonstrates high uptake in certain tumour types.

### 2.2.10. PSMA: [$^{68}$Ga]-ligand for the prostate specific membrane antigen

Prostate specific membrane antigen (PSMA), a tumour associated antigen and type II transmembrane protein, is overexpressed on prostate tumour cells and is avidly bound by this tracer.

## 2.3. DEFINITIONS

### 2.3.1. Recommendation criteria for the use of PET

Based on the review of the literature, the use of PET for clinical indications is classified as recommended, potentially recommended, possibly recommended, or not recommended. These criteria are defined as follows.

*2.3.1.1. Recommended (all conditions below need to be met)*

(a) There is evidence of improved diagnostic performance (higher sensitivity and specificity) compared with other current techniques.
(b) The information derived from PET influences clinical practice.
(c) The information derived from PET has a plausible impact on the patient's outcome, either through adoption of more effective therapeutic strategies or through non-adoption of futile, ineffective or harmful practices.

*2.3.1.2. Potentially recommended (potentially useful)*

There is evidence of improved diagnostic performance (greater sensitivity and specificity) compared with other current techniques, but evidence of an impact on treatment and outcome is lacking.

*2.3.1.3. Possibly recommended (appropriateness not yet documented)*

There is insufficient evidence for assessment, although there is a strong rationale for an expected clinical benefit from PET.

*2.3.1.4. Not recommended*

Improved accuracy of tumour staging will not alter management, or the performance of PET is poorer than that of other current techniques.

**2.3.2. Indications for PET**

Different indications for PET are considered here:

(a) Diagnosis;
(b) Staging;
(c) Monitoring response to multimodality therapy;
(d) Detection/restaging of suspected recurrence;
(e) Follow-up (i.e. planned at regular intervals post-treatment in the absence of any signs/symptoms);
(f) Radiotherapy planning.

When relevant, refinements to these definitions are made in the respective tumour type sections [4].

2.4. SUMMARY OF RESULTS

While only cancer types for which PET has an established role have been included in this publication, it is not intended to be exhaustive in covering all possible pathologies in which PET may add value to management. Cancers in the following organ systems have been considered [3]:

(a) Central nervous system:
   (i) Primary tumours of the central nervous system.
(b) Head and neck:
   (i) Head and neck cancer.
(c) Thoracic:
   (i) Non-small cell lung cancer;
   (ii) Small cell lung cancer;
   (iii) Mesothelioma.

(d)  Breast:
  (i)  Breast cancer.
(e)  Gastrointestinal:
  (i)  Oesophageal cancer;
  (ii)  Gastric cancer;
  (iii)  Colorectal cancer;
  (iv)  Anal cancer;
  (v)  Pancreatic adenocarcinoma;
  (vi)  Hepatocellular carcinoma;
  (vii)  Cholangio- and gall bladder carcinomas.
(f)  Genitourinary:
  (i)  Renal cancer;
  (ii)  Urothelial and bladder cancers;
  (iii)  Germinal cancer;
  (iv)  Prostate cancer.
(g)  Gynaecological:
  (i)  Ovarian cancer;
  (ii)  Endometrial cancer;
  (iii)  Cervical cancer;
  (iv)  Vulvar cancer.
(h)  Bone and soft tissue:
  (i)  Bone and soft tissue sarcomas.
(i)  Cutaneous:
  (i)  Melanoma.
(j)  Haematological:
  (i)  Lymphomas;
  (ii)  Myeloma.
(k)  Endocrine:
  (i)  Thyroid cancer;
  (ii)  Adrenal cancer.
(l)  Neuroendocrine:
  (i)  Bronchial carcinoid;
  (ii)  Gastroenteropancreatic neuroendocrine tumours (GEP–NETs);
  (iii)  Paraganglioma, pheochromocytoma and neuroblastoma;
  (iv)  Medullary thyroid carcinoma.
(m)  Unknown primary:
  (i)  Cancer of unknown primary.

Table 1 summarizes the current evidence for PET in the conditions and indications covered by this publication. Recommendations in several cases refer to radiotracers other than FDG. For full details see the relevant sections.

TABLE 1. INDICATIONS FOR PET

| Type of cancer | Recommendation | Diagnosis | Staging | Response evaluation | Recurrence | Follow-up | RT planning |
|---|---|---|---|---|---|---|---|
| Primary tumours of the central nervous system | Recommended | | | | | | |
| | Potentially recommended | | | | ✓ | | |
| | Possibly recommended | | | ✓ | | | ✓ |
| | Not recommended | ✓ | ✓ | | ✓ | ✓ | |
| Head and neck cancer | Recommended | | | ✓ | ✓ | | ✓ |
| | Potentially recommended | | ✓ | | | | |
| | Possibly recommended | | | | | | |
| | Not recommended | ✓ | | | | ✓ | |

TABLE 1. INDICATIONS FOR PET (cont.)

| Type of cancer | Recommendation | Diagnosis | Staging | Response evaluation | Recurrence | Follow-up | RT planning |
|---|---|---|---|---|---|---|---|
| Non-small cell lung cancer[a] (NSCLC) | Recommended | ✓ | ✓ | ✓ | ✓ | | |
| | Potentially recommended | | | | | | ✓ |
| | Possibly recommended | | | | | | |
| | Not recommended | | | | | ✓ | |
| Small cell lung cancer (SCLC)[a] | Recommended | ✓ | | | | | |
| | Potentially recommended | | | | | | |
| | Possibly recommended | | ✓ | ✓ | ✓ | | ✓ |
| | Not recommended | | | | | ✓ | |

## TABLE 1. INDICATIONS FOR PET (cont.)

| Type of cancer | Recommendation | Diagnosis | Staging | Response evaluation | Recurrence | Follow-up | RT planning |
|---|---|---|---|---|---|---|---|
| Mesothelioma | Recommended | | | | | | |
| | Potentially recommended | | | | | | ✓ |
| | Possibly recommended | | ✓ | ✓ | | | |
| | Not recommended | ✓ | | | ✓ | ✓ | |
| Breast cancer | Recommended | | ✓ | | ✓ | | |
| | Potentially recommended | | | ✓ | | | |
| | Possibly recommended | | | | | ✓ | ✓ |
| | Not recommended | | | | | | |

TABLE 1. INDICATIONS FOR PET (cont.)

| Type of cancer | Recommendation | Diagnosis | Staging | Response evaluation | Recurrence | Follow-up | RT planning |
|---|---|---|---|---|---|---|---|
| Oesophageal cancer | Recommended | | ✓ | ✓ | ✓ | | |
| | Potentially recommended | | | | | | ✓ |
| | Possibly recommended | | | | | | |
| | Not recommended | ✓ | | | | ✓ | |
| Gastric cancer | Recommended | | | | | | |
| | Potentially recommended | | | | | | |
| | Possibly recommended | | ✓ | ✓ | | | |
| | Not recommended | ✓ | | | ✓ | ✓ | ✓ |

10

## TABLE 1. INDICATIONS FOR PET (cont.)

| Type of cancer | Recommendation | Diagnosis | Staging | Response evaluation | Recurrence | Follow-up | RT planning |
|---|---|---|---|---|---|---|---|
| Colorectal cancer | Recommended | | | | ✓ | | |
| | Potentially recommended | | | | | | |
| | Possibly recommended | | ✓ | ✓ | | ✓ | ✓ |
| | Not recommended | ✓ | | | | | |
| Anal cancer | Recommended | | | | ✓ | | ✓ |
| | Potentially recommended | | ✓ | | | | |
| | Possibly recommended | | | | | | |
| | Not recommended | ✓ | | | | ✓ | |

TABLE 1. INDICATIONS FOR PET (cont.)

| Type of cancer | Recommendation | Diagnosis | Staging | Response evaluation | Recurrence | Follow-up | RT planning |
|---|---|---|---|---|---|---|---|
| Pancreatic adenocarcinoma | Recommended | | | | | | |
| | Potentially recommended | | | | ✓ | | |
| | Possibly recommended | | ✓ | ✓ | | | ✓ |
| | Not recommended | ✓ | | | | ✓ | |
| Hepatocellular carcinoma | Recommended | | | | | | |
| | Potentially recommended | | | | | | |
| | Possibly recommended | ✓ | | ✓ | | | |
| | Not recommended | | ✓ | | ✓ | ✓ | ✓ |

TABLE 1. INDICATIONS FOR PET (cont.)

| Type of cancer | Recommendation | Diagnosis | Staging | Response evaluation | Recurrence | Follow-up | RT planning |
|---|---|---|---|---|---|---|---|
| Cholangio- and gall bladder carcinomas | Recommended | | | | | | |
| | Potentially recommended | | | | | | |
| | Possibly recommended | | ✓ | | | | |
| | Not recommended | ✓ | | ✓ | ✓ | ✓ | ✓ |
| Urothelial and bladder cancers | Recommended | | | | | | |
| | Potentially recommended | | | | | | |
| | Possibly recommended | | ✓ | ✓ | ✓ | | |
| | Not recommended | ✓ | | | | ✓ | ✓ |

TABLE 1. INDICATIONS FOR PET (cont.)

| Type of cancer | Recommendation | Diagnosis | Staging | Response evaluation | Recurrence | Follow-up | RT planning |
|---|---|---|---|---|---|---|---|
| Germinal tumours | Recommended | | | ✓ | | | |
| | Potentially recommended | | | | | | |
| | Possibly recommended | | | | ✓ | | |
| | Not recommended | ✓ | ✓ | | | ✓ | ✓ |
| Prostate cancer | Recommended | | | | ✓ | | |
| | Potentially recommended | | ✓ | | | | |
| | Possibly recommended | | | ✓ | | | |
| | Not recommended | ✓ | | | | ✓ | ✓ |

TABLE 1. INDICATIONS FOR PET (cont.)

| Type of cancer | Recommendation | Diagnosis | Staging | Response evaluation | Recurrence | Follow-up | RT planning |
|---|---|---|---|---|---|---|---|
| Ovarian cancer | Recommended | | | | ✓ | | |
| | Potentially recommended | | | | | | |
| | Possibly recommended | | ✓ | ✓ | | | |
| | Not recommended | ✓ | | | | ✓ | ✓ |
| Endometrial cancer | Recommended | | | | ✓ | | |
| | Potentially recommended | | | | | | |
| | Possibly recommended | | ✓ | ✓ | | | |
| | Not recommended | ✓ | | | | ✓ | ✓ |

## TABLE 1. INDICATIONS FOR PET (cont.)

| Type of cancer | Recommendation | Diagnosis | Staging | Response evaluation | Recurrence | Follow-up | RT planning |
|---|---|---|---|---|---|---|---|
| Cervical cancer | Recommended | | ✓ | ✓ | ✓ | | ✓ |
| | Potentially recommended | | | | | | |
| | Possibly recommended | | | | | | |
| | Not recommended | ✓ | | | ✓ | ✓ | |
| Vulvar cancer | Recommended | | ✓ | | | | |
| | Potentially recommended | | | | ✓ | | |
| | Possibly recommended | | | ✓ | | | |
| | Not recommended | ✓ | | | | ✓ | ✓ |

TABLE 1. INDICATIONS FOR PET (cont.)

| Type of cancer | Recommendation | Diagnosis | Staging | Response evaluation | Recurrence | Follow-up | RT planning |
|---|---|---|---|---|---|---|---|
| Bone and soft tissue sarcoma | Recommended | | | | | | |
| | Potentially recommended | | | ✓ | ✓ | | |
| | Possibly recommended | ✓ | ✓ | | | | |
| | Not recommended | | | | | ✓ | ✓ |
| Lymphoma | Recommended | | ✓ | ✓ | ✓ | | |
| | Potentially recommended | | | | | | |
| | Possibly recommended | ✓ | | | | | ✓ |
| | Not recommended | | | | | ✓ | |

17

TABLE 1. INDICATIONS FOR PET (cont.)

| Type of cancer | Recommendation | Diagnosis | Staging | Response evaluation | Recurrence | Follow-up | RT planning |
|---|---|---|---|---|---|---|---|
| Melanoma[b] | Recommended | | | | | | |
| | Potentially recommended | | ✓ | ✓ | ✓ | | |
| | Possibly recommended | | | | | ✓ | |
| | Not recommended | | | | | | ✓ |
| Myeloma | Recommended | | | | | | |
| | Potentially recommended | ✓ | ✓ | ✓ | ✓ | | |
| | Possibly recommended | | | | | | |
| | Not recommended | | | | | ✓ | ✓ |

TABLE 1. INDICATIONS FOR PET (cont.)

| Type of cancer | Recommendation | Diagnosis | Staging | Response evaluation | Recurrence | Follow-up | RT planning |
|---|---|---|---|---|---|---|---|
| Thyroid cancer | Recommended | | | | ✓ | | |
| | Potentially recommended | | | | | | |
| | Possibly recommended | | ✓ | | | | |
| | Not recommended | ✓ | | ✓ | | ✓ | ✓ |
| Adrenocortical cancer | Recommended | | | | | | |
| | Potentially recommended | | ✓ | | | | |
| | Possibly recommended | | | ✓ | ✓ | | |
| | Not recommended | ✓ | | | | ✓ | ✓ |

TABLE 1. INDICATIONS FOR PET (cont.)

| Type of cancer | Recommendation | Diagnosis | Staging | Response evaluation | Recurrence | Follow-up | RT planning |
|---|---|---|---|---|---|---|---|
| Bronchial carcinoid | Recommended | | | | | ✓ | |
| | Potentially recommended | ✓ | | | | | |
| | Possibly recommended | | ✓ | ✓ | ✓ | | |
| | Not recommended | ✓ | | | | | ✓ |
| Gastrointestinal and pancreatic (GEP) neuroendocrine tumour (NET) | Recommended | | ✓ | | ✓ | ✓ | |
| | Potentially recommended | ✓ | | ✓ | | | |
| | Possibly recommended | | | | | | |
| | Not recommended | | | | | | ✓ |

TABLE 1. INDICATIONS FOR PET (cont.)

| Type of cancer | Recommendation | Diagnosis | Staging | Response evaluation | Recurrence | Follow-up | RT planning |
|---|---|---|---|---|---|---|---|
| Paraganglioma, pheochromocytoma and neuroblastoma | Recommended | | ✓ | ✓ | | | |
| | Potentially recommended | | | | | ✓ | |
| | Possibly recommended | ✓ | | | ✓ | | |
| | Not recommended | | | ✓ | | | ✓ |
| Medullary thyroid carcinoma | Recommended | | | | | | |
| | Potentially recommended | | | | | | |
| | Possibly recommended | | | | ✓ | | |
| | Not recommended | | | ✓ | | | |
| Cancer of unknown primary | Recommended | ✓ | ✓ | ✓ | | ✓ | ✓ |

TABLE 1. INDICATIONS FOR PET (cont.)

| Type of cancer | Recommendation | Diagnosis | Staging | Response evaluation | Recurrence | Follow-up | RT planning |
|---|---|---|---|---|---|---|---|
| | Potentially recommended | | | | | | |
| | Possibly recommended | | ✓ | ✓ | ✓ | | |
| | Not recommended | | | | | ✓ | ✓ |

[a] In diagnosis, the indication is only valid for evaluation of the solitary pulmonary nodule.
[b] Not recommended in low risk patients.

# 3. PRIMARY TUMOURS OF THE CENTRAL NERVOUS SYSTEM

Useful PET tracers for central nervous system tumours include FDG and amino acid tracers such as acetate, methionine, FET and DOPA. Amino acid tracers are preferable because their distribution is characterized by lower background activity in the brain. Final choice of the tracer depends on local availability [6].

## 3.1. DIAGNOSIS

*Recommendation: Not recommended*

There is insufficient evidence for the use of PET in this indication. In addition, the correlation between PET tracer uptake and tumour grade is insufficient to support a role for PET in tumour grading [6].

## 3.2. STAGING

*Recommendation: Not recommended*

There are currently insufficient data to support a role for PET in this indication. Magnetic resonance imaging (MRI) provides excellent anatomic definition to determine the local extent of the tumour [6] and distant metastases from primary central nervous system tumours are exceedingly rare.

## 3.3. RESPONSE EVALUATION

*Recommendation: Possibly recommended*

Although a strong rationale exists for this indication, there are few reports regarding the use of PET tracers to assess tumour response to multimodality therapy [7].

## 3.4. RECURRENCE

### 3.4.1. Suspected recurrence

*Recommendation: Possibly recommended*

PET may provide additional information to that provided by MRI or CT for the detection of recurrence following resection or radiotherapy and may identify the region most likely to yield a representative biopsy.

### 3.4.2. Confirmed recurrence

*Recommendation: Not recommended*

There is generally no requirement to further define the tumour using PET when recurrence has been confirmed by other imaging modalities.

## 3.5. FOLLOW-UP

*Recommendation: Not recommended*

There are currently insufficient data to support a role for PET. MRI is adequate for this indication [8].

## 3.6. RADIATION THERAPY PLANNING

*Recommendation: Possibly recommended*

PET currently has no role in defining radiation fields or doses. However, there is a rationale for using PET for dose escalation to the metabolically intense region within the tumour [8].

**Note:** For central nervous system lymphoma, see the discussion on lymphomas in Section 26.

# 4. HEAD AND NECK CANCERS

The following discussion does not include thyroid cancer, which is considered in Section 28.

## 4.1. DIAGNOSIS

*Recommendation: Not recommended*

There are currently insufficient data to support a role for FDG PET in this indication. The diagnosis of primary head and neck cancers is based on clinical examination, endoscopy with biopsies and imaging with CT, MRI and/or ultrasound. See Section 34 for cervical lymph node metastasis of tumours from unknown origin [9].

## 4.2. STAGING

*Recommendation: Potentially recommended*

Use of CT or MRI remains the standard of care for tumoural and nodal staging in this setting. FDG PET is accurate in detecting regional nodal disease, distant metastases and synchronous tumours [6].

## 4.3. RESPONSE EVALUATION

*Recommendation: Recommended*

If performed at least 12 weeks after multimodality treatment, FDG PET is accurate in detecting residual disease. If performed earlier, false positive results due to inflammatory changes are possible. Persistently enlarged FDG negative lymph nodes require close clinical monitoring [6].

## 4.4. RECURRENCE

*Recommendation: Recommended*

Since scarring with distortion of tissue structures following surgery and radiation therapy may limit the diagnostic performance of anatomic imaging techniques, the use of FDG PET in both suspected and confirmed recurrence is recommended [6].

## 4.5. FOLLOW-UP

*Recommendation: Not recommended*

There is insufficient evidence that FDG PET is useful in asymptomatic patients after presumed complete remission [6].

## 4.6. RADIATION THERAPY PLANNING

*Recommendation: Recommended*

Data demonstrate that target volumes and doses may be modified on the basis of FDG PET findings. FDG PET is helpful for the inclusion or exclusion of lymph nodes in the radiation field, although no data on patient outcome are available [10].

# 5. NON-SMALL CELL LUNG CANCER (NSCLC)

## 5.1. DIAGNOSIS (CHARACTERIZATION OF SOLITARY PULMONARY NODULES)

*Recommendation: Recommended*

Solitary pulmonary nodules (SPNs) are common and present a diagnostic challenge, particularly in persons with chronic pulmonary disease or any other condition where biopsy may be risky. FDG PET may be used to stratify SPNs 1 cm or larger as being of high or low risk for malignancy. SPNs with high

FDG uptake ought to be considered malignant until proven otherwise, whereas lesions with low or no uptake ought to be considered for surveillance using CT scanning. The use of PET for characterization of SPNs is cost effective [11].

## 5.2. STAGING

*Recommendation: Recommended*

The use of FDG PET represents the standard of care for staging non-small cell lung cancer (NSCLC), with meta-analyses indicating a higher sensitivity and specificity for PET than for CT scanning. PET is especially valuable in the detection of mediastinal lymph node metastases that are equivocal or negative on CT. In this context, the role of FDG PET is in identifying nodal lesions requiring biopsy for confirmation [12]. Brain metastases are not detected adequately using FDG PET.

## 5.3. RESPONSE EVALUATION

*Recommendation: Recommended*

After multimodality treatment, FDG PET response may be used to select candidates for subsequent potentially life extending therapies. Survival following multimodality treatment is strongly predicted by FDG PET, with improved survival in patients whose tumours show no uptake on post-treatment PET. This predictive value is much greater than that based on CT response [13].

## 5.4. RECURRENCE

*Recommendation: Recommended*

The value of FDG PET in proven or suspected recurrence has been confirmed, allowing the selection of appropriate therapy [13].

## 5.5. FOLLOW-UP

*Recommendation: Not recommended*

While recurrence can probably be detected earlier by PET than by clinical examination or by other imaging modalities, there is no evidence that patient management or survival would be affected [14].

## 5.6. RADIATION THERAPY PLANNING

*Recommendation: Potentially recommended*

The information provided by FDG PET alters the size of radiation therapy treatment fields in over one third of the cases. In most cases, the field size is increased to incorporate FDG positive areas, while in some cases the field size is reduced to avoid unnecessary radiation to adjacent normal tissues, especially in the proximity of critical anatomical structures [14].

# 6. SMALL CELL LUNG CANCER (SCLC)

## 6.1. DIAGNOSIS

*Recommendation: Recommended*

See Section 5.1, as well as Ref. [11].

## 6.2. STAGING

*Recommendation: Possibly recommended*

Management of SCLC is based on staging derived predominantly from CT findings, although FDG PET may be used to confirm limited stage disease. MRI or CT ought to be used to assess metastasis to the brain [15].

## 6.3. RESPONSE EVALUATION

*Recommendation: Not recommended*

As SCLC shrinks rapidly in response to effective treatment, it is unlikely that FDG PET would contribute significantly to the assessment of treatment response [16].

## 6.4. SUSPECTED RECURRENCE

*Recommendation: Not recommended*

The high FDG uptake of SCLC suggests that PET is a sensitive tool for identifying recurrence, although there is no evidence that PET in this context would alter clinical management [17].

## 6.5. FOLLOW-UP

*Recommendation: Not recommended*

Recurrence of SCLC is considered not to be amenable to potentially curative treatments, with CT providing adequate detection of recurrence [17].

## 6.6. RADIATION THERAPY PLANNING

*Recommendation: Possibly recommended*

It is likely that FDG PET would have the same benefit for SCLC as has been demonstrated for NSCLC, resulting in a modification of the radiation therapy field definition in a high proportion of cases [16].

# 7. MESOTHELIOMA

## 7.1. DIAGNOSIS

*Recommendation: Not recommended*

There are currently insufficient data to support a role for FDG PET in this indication, as diagnosis is based on pleural biopsies [18].

## 7.2. STAGING

*Recommendation: Possibly recommended*

Staging of patients with malignant pleural mesothelioma is to assess whether they are candidates for surgical resection. Contrast enhanced CT of the chest and abdomen is the standard for imaging for this purpose. Nonetheless, FDG PET–CT may be useful in identifying distant metastases. When utilized, PET ought to be performed prior to pleurodesis [18].

## 7.3. RESPONSE EVALUATION

*Recommendation: Possibly recommended*

FDG PET may play a role in assessing response to multimodality therapy [19]. Its utility following talc pleurodesis, however, is limited.

## 7.4. RECURRENCE

*Recommendation: Not recommended*

Although the high FDG uptake suggests that PET is a sensitive tool for identifying recurrence, there is no evidence that FDG PET alters clinical management [18].

## 7.5.  FOLLOW-UP

*Recommendation: Not recommended*

Recurrence is considered not to be amenable to potentially curative treatments, with CT providing adequate detection of recurrence [20].

## 7.6.  RADIATION THERAPY PLANNING

*Recommendation: Possibly recommended*

There are several reports that FDG PET influences target volume delineation in intensity modulated radiation therapy [21].

# 8.  BREAST CANCER

## 8.1.  DIAGNOSIS

*Recommendation: Not recommended*

The uptake of FDG in primary breast cancers is related to tumour size, histology and grade. Multiple prospective studies have shown a low sensitivity (25%) for primary tumours 1 cm or smaller in diameter. There is good correlation between FES uptake and oestrogen receptor expression [22].

## 8.2.  STAGING

*Recommendation: Potentially recommended*

FDG PET allows the detection of extra-axillary nodes and distant metastases with higher sensitivity than other diagnostic imaging methods. An exception is brain metastases, where MRI is the method of choice. The relative role of bone scans using $^{99m}$Tc labelled compounds or FDG PET in the detection of bone metastases remains undefined. Nevertheless, bone metastases from breast cancer tend to be osteolytic, and such lesions are known to be detected by FDG PET with higher sensitivity than sclerotic bone metastases. The negative predictive value of FDG PET is too low to stage the axilla reliably, as micrometastases may be missed. FDG PET cannot replace sentinel node biopsy [23]. The sensitivity of FDG PET may be suboptimal in lobular carcinoma [24].

## 8.3. RESPONSE EVALUATION

*Recommendation: Possibly recommended*

FDG PET is most helpful when the results of standard imaging are equivocal or suspicious [25].

## 8.4. RECURRENCE

*Recommendation: Potentially recommended*

Due to its high sensitivity for distant metastases, particularly lymph node and skeletal metastases, FDG PET is helpful in establishing the extent of recurrent disease. There is a role for FDG PET in the detection of suspected recurrence, especially in patients with rising tumour associated markers. So far, however, prospective trials that also address the issues of management changes, outcome and cost efficiency are lacking [25].

## 8.5. FOLLOW-UP

*Recommendation: Not recommended*

There are currently insufficient data to support a role for FDG or FES PET in this indication, including patients on long term therapy [22].

## 8.6. RADIATION THERAPY PLANNING

*Recommendation: Not recommended*

There are currently insufficient data to support a role for FDG PET in this indication [22].

# 9. OESOPHAGEAL CANCER

## 9.1. DIAGNOSIS

*Recommendation: Not recommended*

There is currently no evidence that the addition of FDG PET improves the diagnostic accuracy of endoscopic ultrasound and biopsy [26].

## 9.2. STAGING

*Recommendation: Recommended*

FDG PET is the imaging modality of choice for pre-therapy assessment of disease, especially when the presence of metastatic disease is unknown. In this context, PET may be especially valuable in preventing futile surgery or guiding multimodality therapy [27].

## 9.3. RESPONSE EVALUATION

*Recommendation: Recommended*

FDG PET identifies locoregional disease that is unresponsive to neoadjuvant therapy and interval metastases prior to planned surgery. The endoscopic findings ought to be taken into consideration, as oesophagitis may mimic residual disease on PET [28].

## 9.4. RECURRENCE

*Recommendation: Recommended*

This recommendation is particularly relevant for lower stage tumours treated with local techniques that have recurred locally and remain amenable to potentially curative locoregional therapy [29].

## 9.5. FOLLOW-UP

*Recommendation: Not recommended*

There are no data indicating a role for FDG PET in follow-up of oesophageal cancer [29].


## 9.6. RADIATION THERAPY PLANNING

*Recommendation: Potentially recommended*

There is a strong rationale for the use of FDG PET in radiation therapy planning, although clinical outcome data are lacking. Whenever available, 4-D CT, or other motion management methods, ought to be employed to allow better delineation of target volumes [26].


# 10. GASTRIC CANCER

The following discussion refers to distal gastric cancers. Tumours involving the gastroesophageal junction are generally considered to be distal oesophageal carcinomas.


## 10.1. DIAGNOSIS

*Recommendation: Not recommended*

There is currently no evidence that the addition of FDG PET to endoscopy and biopsy improves diagnostic performance. The normal gastric mucosa shows some level of physiological FDG uptake, as do several benign conditions [15].

## 10.2. STAGING

*Recommendation: Possibly recommended*

There are limited data on the value of FDG PET in detecting locoregional lymph node and distant metastatic disease [27].

## 10.3. RESPONSE EVALUATION

*Recommendation: Possibly recommended*

FDG PET may identify response to neoadjuvant therapy. There are, however, no data to determine the impact of PET on clinical outcome [30].

## 10.4. RECURRENCE

*Recommendation: Not recommended*

There are currently insufficient data to support a role for FDG PET in this indication [31].

## 10.5. FOLLOW-UP

*Recommendation: Not recommended*

There are currently insufficient data to support a role for FDG PET in this indication [31].

## 10.6. RADIATION THERAPY PLANNING

*Recommendation: Not recommended*

There are currently insufficient data to support a role for FDG PET in this indication. Palliative radiation therapy is targeted at the CT defined mass; curative post-operative radiation therapy (usually associated with chemotherapy) is targeted at the surgical bed.

**Note:** FDG PET may have reduced sensitivity in mucinous and diffuse adenocarcinomas [30].

# 11. COLORECTAL CANCER

## 11.1. DIAGNOSIS

*Recommendation: Not recommended*

Any symptoms suggestive of colorectal cancer need to be investigated by endoscopy, with biopsy of suspicious lesions. Incidental focal FDG avid colorectal lesions are associated with a high risk of pre-malignancy or malignancy and ideally ought to be investigated further [32].

## 11.2. STAGING

*Recommendation: Possibly recommended*

FDG PET has not shown superiority to CT and/or MRI in T or N staging, or in the evaluation of hepatic metastases [33]. In high risk, therapy naive patients being considered for hepatic resection, FDG PET may be superior to other imaging modalities for the detection of extrahepatic metastases [34]. In the context of recent neoadjuvant therapy, FDG PET–CT is less reliable than CT and MRI.

## 11.3. RESPONSE EVALUATION

*Recommendation: Possibly recommended*

FDG PET may be useful in excluding local disease progression in the liver after ablative therapy (superior to CT, complementary to MRI) [35]. In other contexts, FDG PET has little or no additional benefit to CT or MRI assessment [36].

## 11.4. RECURRENCE

*Recommendation: Recommended*

FDG PET is valuable in the evaluation of patients with suspected recurrence (e.g. due to elevated carcinoembryonic antigen levels). Evidence suggests that in this context FDG PET is more sensitive than CT [36].

## 11.5. FOLLOW-UP

*Recommendation: Possibly recommended*

In patients with contraindications for iodine contrast medium, FDG PET may be considered as an alternative to CT [37].

## 11.6. RADIATION THERAPY PLANNING

*Recommendation: Possibly recommended*

In rectal cancer there are limited data to support the use of FDG PET in assisting with the definition of radiation fields.

**Note:** Although mucinous adenocarcinomas are usually not FDG avid, there is in fact evidence that FDG uptake by rectal cancer is similar in mucinous and nonmucinous histological subtypes [38].

# 12. ANAL CANCER

## 12.1. DIAGNOSIS

*Recommendation: Not recommended*

There are currently insufficient data to support a role for FDG PET in this indication [39].

## 12.2. STAGING

*Recommendation: Potentially recommended*

In high risk patients, FDG PET is superior to other imaging modalities for detecting locoregional lymph node and distant metastases [39].

## 12.3. RESPONSE EVALUATION

*Recommendation: Recommended*

In high risk patients for whom salvage surgery is considered, FDG PET provides a sensitive assessment of response to chemoradiation [39].

## 12.4. RECURRENCE

*Recommendation: Recommended*

The exam is required in high risk patients for whom salvage surgery is considered [40].

## 12.5. FOLLOW-UP

*Recommendation: Not recommended*

There are currently insufficient data to support a role for FDG PET in this indication [41].

## 12.6. RADIATION THERAPY PLANNING

*Recommendation: Recommended*

For anal cancer there are data to support the use of FDG PET in assisting with the placement of radiation fields [41].

# 13. PANCREATIC ADENOCARCINOMA

## 13.1. DIAGNOSIS

*Recommendation: Not recommended*

There are currently insufficient data to support a role for FDG PET in this indication: FDG PET cannot replace diagnostic CT and ultrasound with biopsy [42].

## 13.2. STAGING

*Recommendation: Possibly recommended*

FDG PET may be recommended for high risk patients who are candidates for therapies with curative intent. For metastatic staging, FDG PET may complement conventional imaging modalities [31].

## 13.3. RESPONSE EVALUATION

*Recommendation: Possibly recommended*

There is a rationale for the use of FDG PET for the assessment of response to systemic therapy [43].

## 13.4. RECURRENCE

*Recommendation: Potentially recommended*

The degree of FDG avidity may help distinguish recurrence from post-treatment changes [44].

## 13.5. FOLLOW-UP

*Recommendation: Not recommended*

There are currently insufficient data to support a role for FDG PET in this indication [45].

## 13.6. RADIATION THERAPY PLANNING

*Recommendation: Possibly recommended*

FDG PET data may be useful for target volume delineation and dose intensification [45].

# 14. HEPATOCELLULAR CARCINOMA

## 14.1. DIAGNOSIS

*Recommendation: Possibly recommended*

FDG PET alone has no role in diagnosis. Several studies have suggested its usefulness in combination with acetate to detect well differentiated tumours. A similar role has been described for choline [46].

## 14.2. STAGING

*Recommendation: Not recommended*

There are currently insufficient data to support a role for PET in this indication [47].

## 14.3. RESPONSE EVALUATION

*Recommendation: Possibly recommended*

Limited data suggest that persistent FDG uptake after chemotherapy is a negative prognostic feature in patients considered for liver transplantation [48].

## 14.4. RECURRENCE

*Recommendation: Not recommended*

There are currently insufficient data to support a role for PET in this indication [49].

## 14.5. FOLLOW-UP

*Recommendation: Not recommended*

There are currently insufficient data to support a role for FDG PET in this indication [47].

## 14.6. RADIATION THERAPY PLANNING

*Recommendation: Not recommended*

There are currently insufficient data to support a role for FDG PET in this indication [47].

# 15.  CHOLANGIO- AND GALLBLADDER CARCINOMAS

## 15.1. DIAGNOSIS

*Recommendation: Not recommended*

There are currently insufficient data to support a role for FDG PET alone in this indication. Several studies have suggested its usefulness in combination with acetate to detect well differentiated tumours. A similar role has been described for choline [50].

## 15.2. STAGING

*Recommendation: Potentially recommended*

In limited series, FDG PET is more accurate than CT scanning for identifying extrahepatic lesions in patients being considered for surgical resection [50].

## 15.3. RESPONSE EVALUATION

*Recommendation: Not recommended*

There are currently insufficient data to support a role for FDG PET in this indication [51].

## 15.4. RECURRENCE

*Recommendation: Not recommended*

There are currently insufficient data to support a role for FDG PET in this indication [51].

## 15.5. FOLLOW-UP

*Recommendation: Not recommended*

There are currently insufficient data to support a role for FDG PET in this indication [52].

## 15.6. RADIATION THERAPY PLANNING

*Recommendation: Not recommended*

There are currently insufficient data to support a role for FDG PET in this indication [52].

# 16. RENAL CANCER

While FDG PET is sensitive in the detection of extrarenal metastasis in renal cell carcinoma, there is currently insufficient evidence for the value of FDG PET in the management of this cancer [53].

# 17. UROTHELIAL AND BLADDER CANCERS

## 17.1. DIAGNOSIS

*Recommendation: Not recommended*

There are currently insufficient data to support a role for FDG PET in this indication [54].

## 17.2. STAGING

*Recommendation: Possibly recommended*

FDG PET may be useful in patients with muscle invasive or suspected metastatic bladder cancer. There are no data supporting the use of FDG PET for urethral and ureteral carcinomas, which have a poor detection rate [54].

## 17.3. RESPONSE EVALUATION

*Recommendation: Possibly recommended*

FDG PET may be useful for evaluation of patients with invasive bladder cancer following systemic therapy [55].

## 17.4. RECURRENCE

*Recommendation: Possibly recommended*

In cases of CT findings that are equivocal for metastatic recurrence, FDG PET may be useful to detect recurrence [55].

## 17.5. FOLLOW-UP

*Recommendation: Not recommended*

There are currently insufficient data to support a role for FDG PET in this indication [56].

## 17.6. RADIATION THERAPY PLANNING

*Recommendation: Not recommended*

There are currently insufficient data to support a role for FDG PET in this indication [56].

# 18. GERMINAL TUMOUR

## 18.1. DIAGNOSIS

*Recommendation: Not recommended*

There are currently insufficient data to support a role for FDG PET in this indication [57].

## 18.2. STAGING

*Recommendation: Not recommended*

There are currently insufficient data to support a role for FDG PET in this indication. The negative predictive value is not high enough to avoid adjuvant therapies in the case of negative results [57].

## 18.3. RESPONSE EVALUATION

*Recommendation: Recommended*

FDG PET is superior to CT in patients affected by seminoma. Except for mature teratoma, PET can distinguish residual tumour mass >3 cm from necrosis and/or fibrosis [57].

## 18.4. RECURRENCE

*Recommendation: Possibly recommended*

In cases of equivocal CT findings and/or elevation of serum tumour markers, FDG PET can be used to detect recurrence [58].

## 18.5. FOLLOW-UP

*Recommendation: Not recommended*

There are currently insufficient data to support a role for FDG PET in this indication [58].

## 18.6. RADIATION THERAPY PLANNING

*Recommendation: Not recommended*

Radiation therapy has a minimal role in non-seminomatous germ cell tumours, and there are currently insufficient data to support a role for FDG PET in this indication. For early stage seminomas, for which the patterns of failure are well described, there are no data to suggest that PET may influence radiation fields [57].

# 19. PROSTATE CANCER

The role of FDG PET in patients with prostate cancer is somewhat limited because of the intrinsically low glucose transporter expression of these tumours, except for poorly differentiated types. Other PET tracers such as PSMA ligands and choline are superior for imaging prostate cancer. When available, PSMA ligands (most notably the PSMA-11 ligand) constitute the tracers of choice in this tumour type [59].

## 19.1. DIAGNOSIS

*Recommendation: Not recommended*

There are insufficient data to support a role for PET in this indication. MRI remains the modality of choice after inconclusive biopsy [60].

## 19.2. STAGING

*Recommendation: Potentially recommended*

In high risk patients, PET has superior sensitivity for lymph node and distant metastatic disease as compared to CT, MRI and bone scan. For local staging, there are insufficient data to support PET use [61].

## 19.3. RESPONSE EVALUATION

*Recommendation: Possibly recommended*

Limited data are available to support a role for PET for this indication [61].

## 19.4. RECURRENCE

*Recommendation: Recommended*

The use of PSMA ligand PET for localization of prostate cancer in the setting of biochemical recurrence is especially recommended in patients with low prostate specific antigen (PSA) values (0.2–10 ng/mL) who may be candidates for salvage therapy. For choline PET, PSA values ought to be $\geq 1.0$ ng/mL [60].

## 19.5. FOLLOW-UP

*Recommendation: Not recommended*

There are currently insufficient data to support a role for PET in this indication [62].

## 19.6. RADIATION THERAPY PLANNING

*Recommendation: Not recommended*

There are currently insufficient data to support a role for PET in this indication. Pelvic MRI remains the modality of choice [63].

## 19.7. THERANOSTIC PLANNING

*Recommendation: Recommended*

Documentation of PSMA expression with [68]Ga-PSMA ligand PET is required prior to treatment with radiolabelled PSMA ligand [64].

# 20. OVARIAN CANCER

## 20.1. DIAGNOSIS

*Recommendation: Not recommended*

There are currently insufficient data to support a role for FDG PET in this indication [65].

## 20.2. STAGING

*Recommendation: Possibly recommended*

Staging of ovarian cancers is usually performed surgically with the support of conventional imaging. There is limited evidence that FDG PET may be useful in patients with suspected stage IV disease or when indeterminate lymph nodes are identified on conventional imaging [65].

## 20.3. RESPONSE EVALUATION

*Recommendation: Possibly recommended*

FDG PET may be used for assessing response in patients with known metastatic disease after multimodality therapy [66].

## 20.4. RECURRENCE

*Recommendation: Recommended*

Most studies show that FDG PET is superior to CT and complementary to MRI in patients with suspected recurrence [65].

## 20.5. FOLLOW-UP

*Recommendation: Not recommended*

There are currently insufficient data to support a role for FDG PET in this indication [65].

## 20.6. RADIATION THERAPY PLANNING

*Recommendation: Not recommended*

There are currently insufficient data to support a role for FDG PET in this indication. Radiation therapy has a very limited role in the management of ovarian carcinoma. When used for palliation, radiation therapy is directed at symptomatic masses identified by CT.

**Note:** Mucinous adenocarcinomas may be non-FDG avid, and PET in this particular subgroup may be less sensitive [65].

# 21. ENDOMETRIAL CANCER

## 21.1. DIAGNOSIS

*Recommendation: Not recommended*

There are currently insufficient data to support a role for FDG PET in this indication. The incidental finding of uterine activity in post-menopausal patients requires further workup to exclude uterine malignancy [67].

## 21.2. STAGING

*Recommendation: Potentially recommended*

In advanced disease FDG PET may be recommended to exclude distant metastasis [67].

## 21.3. RESPONSE EVALUATION

*Recommendation: Not recommended*

There are currently insufficient data to support a role for FDG PET in this indication [67].

## 21.4. RECURRENCE

*Recommendation: Potentially recommended*

FDG PET may be used in the workup for patients considered for surgery and/or locoregional therapy [67].

## 21.5. FOLLOW-UP

*Recommendation: Not recommended*

There are currently insufficient data to support a role for FDG PET in this indication [67].

## 21.6. RADIATION THERAPY PLANNING

*Recommendation: Not recommended*

There are currently insufficient data to support a role for FDG PET in this indication [68].

# 22. CERVICAL CANCER

## 22.1. DIAGNOSIS

*Recommendation: Not recommended*

There are currently insufficient data to support a role for FDG PET in this indication [69].

## 22.2. STAGING

*Recommendation: Recommended*

In stage IB–IV cervical cancer, FDG PET is a valuable adjunct to conventional imaging methods, namely CT and MRI. Although MRI is the preferred method for evaluation of local extension, PET is superior for the evaluation of lymph node involvement and sensitive for the detection of distant metastases [69].

## 22.3. RESPONSE EVALUATION

*Recommendation: Recommended*

In high risk cervical cancer, FDG PET ought to be performed 3–6 months after the completion of chemoradiation [66].

## 22.4. RECURRENCE

*Recommendation: Recommended*

In cervical cancer recurrence, FDG PET is recommended for patients where metastasis is suspected [69].

## 22.5. FOLLOW-UP

*Recommendation: Not recommended*

There are currently insufficient data to support a role for FDG PET in this indication [70].

## 22.6. RADIATION THERAPY PLANNING

*Recommendation: Recommended*

For locally advanced tumours, the detection of metastasis in para-aortic lymph nodes by FDG PET may lead to modification of the treatment fields. This is particularly important in cervical cancer [70].

# 23. VULVAR CANCER

## 23.1. DIAGNOSIS

*Recommendation: Not recommended*

There are currently insufficient data to support a role for FDG PET in this indication [71].

## 23.2. STAGING

*Recommendation: Potentially recommended*

In advanced disease FDG PET may be used to exclude distant metastasis [71].

## 23.3. RESPONSE EVALUATION

*Recommendation: Not recommended*

There are currently insufficient data to support a role for FDG PET in this indication [71].

## 23.4. RECURRENCE

*Recommendation: Potentially recommended*

FDG PET may be used in the workup for patients considered for salvage surgery and/or locoregional therapy [71].

## 23.5. FOLLOW-UP

*Recommendation: Not recommended*

There are currently insufficient data to support a role for FDG PET in this indication [71].

## 23.6. RADIATION THERAPY PLANNING

*Recommendation: Not recommended*

There are currently insufficient data to support a role for FDG PET in this indication [71].

# 24. BONE AND SOFT TISSUE SARCOMAS

## 24.1. DIAGNOSIS

*Recommendation: Possibly recommended*

There are limited data to suggest that FDG PET may direct biopsy to identify malignant transformation in pre-existing benign mesenchymal lesions. Gastrointestinal stromal tumours (GISTs) are usually diagnosed by endoscopy and/or biopsy [72].

## 24.2. STAGING

*Recommendation: Possibly recommended*

Sarcomas have a particular propensity for early metastatic spread to the lungs. High resolution CT is more effective than FDG PET for detecting small lung metastases. However, PET may be more useful for detecting extrapulmonary metastases. With GIST patients, a baseline FDG PET may be used to determine tumour avidity for subsequent treatment and response evaluation [73].

## 24.3. RESPONSE EVALUATION

*Recommendation: Potentially recommended*

There are data supporting the use of FDG PET to monitor the response to multimodality therapy. In GIST, FDG avid tumours may be assessed for early response (e.g. after two days) to treatment [73].

## 24.4. RECURRENCE

*Recommendation: Potentially recommended*

Suspected recurrence will usually require biopsy for confirmation, which can be directed by FDG PET [74].

## 24.5. FOLLOW-UP

*Recommendation: Not recommended*

There are currently insufficient data to support a role for FDG PET in this indication [74].

## 24.6. RADIATION THERAPY PLANNING

*Recommendation: Not recommended*

There are currently insufficient data to support a role for FDG PET in this indication.

# 25. MELANOMA

## 25.1. DIAGNOSIS

*Recommendation: Not recommended*

The diagnosis of melanoma requires biopsy and histopathological examination. FDG PET does not reliably distinguish between benign and malignant pigmented skin lesions [75].

## 25.2. STAGING

### 25.2.1. Stages I and II: Low pre-test probability of metastases

*Recommendation: Not recommended*

FDG PET is less sensitive than sentinel node biopsy for staging regional lymph nodes. In patients with low pre-test probability of distant metastases, the sensitivity of FDG PET for distant metastases has been reported to be low. Very small metastases are common in melanoma and may be beyond the resolution of PET, despite the usually high avidity of these tumours for FDG [76].

### 25.2.2. Stages I and II: High pre-test probability of metastases

*Recommendation: Recommended*

In patients with intermediate or high risk of distant metastases (melanoma of the head, neck and trunk, Breslow index >4 mm, ulceration, high mitotic rate), FDG PET is recommended for detecting potentially operable metastases [76].

### 25.2.3. Stage III or potential stage IV

*Recommendation: Potentially recommended*

There is a role for FDG PET in assessing locoregional or distant disease to guide appropriate therapy [76].

## 25.3. RESPONSE EVALUATION

*Recommendation: Potentially recommended*

Recent data support a role for FDG PET in assessing response to immunotargeted therapy [77].

## 25.4. RECURRENCE

*Recommendation: Potentially recommended*

In the case of a suspicious lesion that is not readily amenable to biopsy, high uptake of FDG is strongly suggestive of recurrent melanoma [78].

## 25.5. FOLLOW-UP

*Recommendation: Possibly recommended*

FDG PET may be used as an alternative to CT in high risk patients [79].

## 25.6. RADIATION THERAPY PLANNING

*Recommendation: Not recommended*

There are currently insufficient data to support a role for FDG PET in this indication [75].

**Note:** Squamous and Merkel cell carcinomas will not be discussed in this publication due to the limited role of FDG PET in their treatment [75].

# 26. LYMPHOMA

## 26.1. DIAGNOSIS

*Recommendation: Possibly recommended*

There is a rationale to support the use of FDG PET for the diagnosis of high grade transformation of low grade lymphoma, including chronic lymphocytic leukaemia and/or small lymphocytic lymphoma [80].

## 26.2. STAGING

*Recommendation: Recommended*

FDG PET is the modality of choice for baseline staging but does not necessarily replace the need for contrast enhanced CT. Owing to its superior sensitivity and specificity for most types of lymphomas, FDG PET is recommended for staging of lymphomas known to demonstrate FDG avidity [81]. Bone marrow biopsy is unnecessary in patients with Hodgkin's disease when staged with FDG PET and may be obviated in patients with high grade non-Hodgkin's lymphoma, when positive [82].

## 26.3. RESPONSE EVALUATION

*Recommendation: Recommended*

FDG PET is the method of choice for the assessment of response (including interim PET, when performed) in FDG avid lymphomas. FDG PET has prognostic value in patients who are candidates for bone marrow transplant [80].

## 26.4. RECURRENCE

*Recommendation: Recommended*

FDG PET is the modality of choice for suspected recurrence [83].

## 26.5. FOLLOW-UP

*Recommendation: Not recommended*

There are currently insufficient data to support a role for FDG PET in this indication [83].

## 26.6. RADIATION THERAPY PLANNING

*Recommendation: Possibly recommended*

There are limited data to support the use of FDG PET for radiation therapy planning in patients treated for Hodgkin's disease and non-Hodgkin's lymphoma [84].
**Note:** The above recommendations also apply to primary central nervous system lymphomas [7].


# 27. MYELOMA


## 27.1. DIAGNOSIS

*Recommendation: Potentially recommended*

There is a rationale to support the use of FDG PET to distinguish monoclonal gammopathy of uncertain significance, from plasmacytoma, multiple myeloma and smouldering myeloma [85].


## 27.2. STAGING

*Recommendation: Potentially recommended*

FDG PET is useful for risk stratification to identify FDG avid medullary and extramedullary disease. FDG PET is also useful for confirming solitary skeletal plasmacytoma [86].


## 27.3. RESPONSE EVALUATION

*Recommendation: Potentially recommended*

FDG PET is useful for the assessment of response to therapy. FDG PET has prognostic value in patients who are candidates for bone marrow transplant [85].

## 27.4. RECURRENCE

*Recommendation: Potentially recommended*

FDG PET is useful for suspected recurrence [86].

## 27.5. FOLLOW-UP

*Recommendation: Not recommended*

There are currently insufficient data to support a role for FDG PET in this indication [87].

## 27.6. RADIATION THERAPY PLANNING

*Recommendation: Not recommended*

There are currently insufficient data to support a role for FDG PET in this indication [88].

# 28. THYROID CANCER

## 28.1. DIAGNOSIS

*Recommendation: Not recommended*

There are currently insufficient data to support a role for FDG PET in this indication. FDG avid incidental nodules need to be evaluated with ultrasound guided fine needle aspiration cytology [89].

## 28.2. STAGING

*Recommendation: Possibly recommended*

There is increasing evidence for the use of FDG PET in locally advanced differentiated thyroid cancers. For poorly differentiated or anaplastic cancers, the evidence supporting the use of FDG PET is more established [90].

## 28.3. RESPONSE EVALUATION

*Recommendation: Not recommended*

There are currently insufficient data to support a role for FDG PET in this indication, except for poorly differentiated and anaplastic cancers [91].

## 28.4. RESTAGING AND SUSPECTED RECURRENCE

*Recommendation: Recommended*

In patients with rising thyroglobulin levels and a negative [131]I iodide whole body scan, FDG PET provides useful data [92].

## 28.5. FOLLOW-UP

*Recommendation: Not recommended*

There are currently insufficient data to support a role for FDG PET in this indication, except for poorly differentiated and anaplastic cancers [93].

## 28.6. RADIATION THERAPY PLANNING

*Recommendation: Not recommended*

There are currently insufficient data to support a role for FDG PET in this indication [94].

# 29.  ADRENOCORTICAL CARCINOMAS

## 29.1. DIAGNOSIS

*Recommendation: Not recommended*

There are currently insufficient data to support a role for FDG PET in this indication [95].

## 29.2. STAGING

*Recommendation: Possibly recommended*

There is evidence for the use of FDG PET in staging adrenocortical carcinomas [95].

## 29.3. RESPONSE EVALUATION

*Recommendation: Possibly recommended*

Limited data are available to support a role for FDG PET in this indication [95].

## 29.4. RECURRENCE

*Recommendation: Possibly recommended*

In patients with equivocal findings on conventional imaging, FDG PET may be considered in identifying recurrence [95].

## 29.5. FOLLOW-UP

*Recommendation: Not recommended*

There are currently insufficient data to support a role for FDG PET in this indication [95].

## 29.6. RADIATION THERAPY PLANNING

*Recommendation: Not recommended*

There are currently insufficient data to support a role for FDG PET in this indication [95].

# 30. BRONCHIAL CARCINOID

In general, [68]Ga-DOTA labelled somatostatin analogues are the tracers of choice for bronchial carcinoid. However, in patients with poorly differentiated forms, FDG may also be considered [96].

## 30.1. DIAGNOSIS

*Recommendation: Not recommended*

There are currently insufficient data to support a role for PET in this indication [97].

## 30.2. STAGING

*Recommendation: Possibly recommended*

In suspected advanced stage disease DOTA-SSA PET shows high sensitivity in detecting metastasis [97].

## 30.3. RESPONSE EVALUATION

*Recommendation: Possibly recommended*

DOTA-SSA may be used to assess treatment response [97].

## 30.4. RECURRENCE

*Recommendation: Possibly recommended*

The high sensitivity of DOTA-SSA allows for the detection of local recurrence and possible metastatic disease [97].

## 30.5. FOLLOW-UP

*Recommendation: Recommended*

Due to the long natural history of these tumours, imaging incorporating DOTA-SSA PET is a sensitive means of detecting disease recurrence and/or progression [97].

## 30.6. RADIATION THERAPY PLANNING

*Recommendation: Not recommended*

There are currently insufficient data to support a role for PET in this indication [97].

## 30.7. THERANOSTIC PLANNING

*Recommendation: Potentially recommended*

While there are limited data for the application of peptide receptor radionuclide therapy (PRRT) in bronchial carcinoid, when such therapy is considered, DOTA-SSA PET is an important part of the patient's workup [98].

# 31. GASTROINTESTINAL AND PANCREATIC NEUROENDOCRINE TUMOURS (GEP-NETs)

In general, [68]Ga-DOTA labelled somatostatin analogues are the tracers of choice for GEP-NETs (except for insulinoma). FDOPA, when available, is an alternative tracer for these tumours. In poorly differentiated NETs, FDG may be recommended [96].

## 31.1. DIAGNOSIS

*Recommendation: Potentially recommended*

In patients in whom there is a high clinical suspicion, DOTA-SSA PET is helpful for identifying disease and assessing surgical resectability [99].

## 31.2. STAGING

*Recommendation: Recommended*

DOTA-SSA PET is the most accurate modality for identifying the true extent of disease. MRI imaging of the liver ought to be considered for accurate assessment of intrahepatic disease burden [99].

## 31.3. RESPONSE EVALUATION

*Recommendation: Potentially recommended*

DOTA-SSA PET may be used to assess treatment response to somatostatin receptor based therapy [99].

## 31.4. RECURRENCE

*Recommendation: Recommended*

The high sensitivity of DOTA-SSA PET allows for the detection of local recurrence and possible metastatic disease [99].

## 31.5. FOLLOW-UP

*Recommendation: Recommended*

Due to the long natural history of these tumours, imaging incorporating DOTA-SSA PET is a sensitive means of detecting disease recurrence and/or progression [99].

## 31.6. RADIATION THERAPY PLANNING

*Recommendation: Not recommended*

There are currently insufficient data to support a role for PET in this indication [99].

## 31.7. THERANOSTIC PLANNING

*Recommendation: Recommended*

DOTA-SSA PET is an important part of the patient's workup, as it identifies potentially suitable candidates for PRRT [98].

# 32. PARAGANGLIOMA, PHAEOCHROMOCYTOMA AND NEUROBLASTOMA

For these neuroectodermal NETs, FDG, FDOPA and DOTA labelled somatostatin analogues may be recommended, depending on the level of differentiation and the clinical context [100].

## 32.1. DIAGNOSIS

*Recommendation: Possibly recommended*

In patients in whom there is a high clinical suspicion, PET may be helpful in identifying disease and surgical resectability [101].

## 32.2. STAGING

*Recommendation: Recommended*

In suspected advanced stage disease PET shows high sensitivity in detecting metastasis. For neuroblastoma, it should be noted that radioiodinated metaiodobenzylguanidine (mIBG) remains the recommended tracer [101].

## 32.3. RESPONSE EVALUATION

*Recommendation: Recommended*

PET is used to assess treatment response. For mIBG avid neuroblastoma, see the note above [101].

## 32.4. RECURRENCE

*Recommendation: Possibly recommended*

The high sensitivity of PET allows for the detection of local recurrence and possible metastatic disease [101].

## 32.5. FOLLOW-UP

*Recommendation: Potentially recommended*

There are not sufficient data to support this application outside of neuroblastoma [101].

## 32.6. RADIATION THERAPY PLANNING

*Recommendation: Not recommended*

There are currently insufficient data to support a role for PET in this indication [101].

## 32.7. THERANOSTIC PLANNING

*Recommendation: Potentially recommended*

While there are limited data for the application of PRRT in these diseases, when such therapy is considered, PET, using the corresponding theranostic agent, is an important part of the patient's workup [102].


# 33. MEDULLARY THYROID CARCINOMA

For medullary thyroid carcinoma, FDG and DOPA are utilized, depending on the clinical context [103].

## 33.1. DIAGNOSIS

*Recommendation: Not recommended*

There are currently insufficient data to support a role for PET in this indication [104].

## 33.2. STAGING

*Recommendation: Not recommended*

There are currently insufficient data to support a role for PET in this indication [104].

## 33.3. RESPONSE EVALUATION

*Recommendation: Not recommended*

There are currently insufficient data to support a role for PET in this indication [104].

## 33.4. RECURRENCE

*Recommendation: Possibly recommended*

While CT, MRI and ultrasound remain the modalities of choice for identifying the site(s) of recurrence, PET has potential utility in patients with rising calcitonin or carcinoembryonic antigen levels [104].

## 33.5. FOLLOW-UP

*Recommendation: Not recommended*

There are currently insufficient data to support a role for PET in this indication [104].

## 33.6. RADIATION THERAPY PLANNING

*Recommendation: Not recommended*

There are currently insufficient data to support a role for PET in this indication [104].

# 34. CANCER OF UNKNOWN PRIMARY

For neuroendocrine tumours of unknown origin, depending on the level of differentiation, either FDG, DOTA-SSA or FDOPA may be recommended.

## 34.1. DIAGNOSIS

### 34.1.1. Cervical adenopathy with occult primary

*Recommendation: Recommended*

FDG PET has a primary tumour detection rate of approximately 31–58% when other tests fail to identify a primary [9].

### 34.1.2. Other metastases of unknown origin

*Recommendation: Possibly recommended*

For raised tumour markers, paraneoplastic syndromes and metastases outside the neck, FDG PET may be used if the conventional workup has failed to identify the primary tumour [105].

## 34.2. STAGING

*Recommendation: Possibly recommended*

FDG PET may be recommended for evaluation of the extent of disease [106].

## 34.3. RESPONSE EVALUATION

*Recommendation: Possibly recommended*

FDG PET may be recommended for evaluation of treatment response where clinically relevant [107].

## 34.4. FOLLOW-UP

*Recommendation: Not recommended*

There are currently insufficient data to support a role for FDG PET in this indication [107].

## 34.5. RADIATION THERAPY PLANNING

*Recommendation: Not recommended*

There are currently insufficient data to support a role for FDG PET in this indication [107].

# REFERENCES

[1]     VERMA, V., et al., Use of PET and other functional imaging to guide target delineation in radiation oncology, Semin. Radiat. Oncol. **28** (2018) 171–177.

[2]     PENNANT, M., et al., A systematic review of positron emission tomography (PET) and positron emission tomography/computed tomography (PET/CT) for the diagnosis of breast cancer recurrence, Health Technol. Assess. **14** 50(2010) 1–103.

[3]     WIBMER, A.G., HRICAK, H., ULANER, G.A., WEBER, W., Trends in oncologic hybrid imaging, Eur. J. Hybrid Imaging **2** (2018) 1.

[4]     POON, R., ONTARIO PET STEERING COMMITTEE (PEBC), CANCER CARE ONTARIO (CCO), PET Six-Month Monitoring Report 2015-2: Evidence from Primary Studies and Systematic Reviews and Recommendations from Clinical Practice Guidelines July to December 2015, PEBC, Toronto (2016).

[5]     INTERNATIONAL ATOMIC ENERGY AGENCY, Appropriate Use of FDG-PET for the Management of Cancer Patients, IAEA Human Health Series No. 9, IAEA, Vienna (2010).

[6]     CASTALDI, P., LECCISOTTI, L., BUSSU, F., MICCICHÈ, F., RUFINI, V., Role of $^{18}$F-FDG PET-CT in head and neck squamous cell carcinoma, Acta Otorhinolaryngol. Ital. **33** (2013) 1–8.

[7]     KAWAI, N., MIYAKE, K., YAMAMOTO, Y., NISHIYAMA, Y., TAMIYA, T., $^{18}$F-FDG PET in the diagnosis and treatment of primary central nervous system lymphoma, Biomed. Res. Int. **2013** (2013) 247152.

[8]     ZHAO, C., ZHANG, Y., WANG, J., A meta-analysis on the diagnostic performance of (18)F-FDG and (11)C-methionine PET for differentiating brain tumors, AJNRAm. J. Neuroradiol. **35** (2014) 1058–1065.

[9]     ZHU, L., WANG, N., $^{18}$F-fluorodeoxyglucose positron emission tomography-computed tomography as a diagnostic tool in patients with cervical nodal metastases of unknown primary site: A meta-analysis, Surg. Oncol. **22** (2013) 190–194.

[10]    TROOST, E.G., SCHINAGL, D.A., BUSSINK, J., OYEN, W.J., KAANDERS, J.H., Clinical evidence on PET–CT for radiation therapy planning in head and neck tumours, Radiother. Oncol. **96** (2010) 328–334.

[11]    FLETCHER, J.W., et al., A comparison of the diagnostic accuracy of $^{18}$F-FDG PET and CT in the characterization of solitary pulmonary nodules, J. Nucl. Med. **49** (2008) 179–185.

[12]    FISCHER, B., et al., Preoperative staging of lung cancer with combined PET-CT, N. Engl. J. Med. **361** (2009) 32–39.

[13]    CREMONESI, M., et al., Role of interim $^{18}$F-FDG–PET/CT for the early prediction of clinical outcomes of non-small cell lung cancer (NSCLC) during radiotherapy or chemo-radiotherapy: A systematic review, Eur. J. Nucl. Med. Mol. Imaging **44** (2017) 1915–1927.

[14] MACHTAY, M., et al., Prediction of survival by [$^{18}$F]fluorodeoxyglucose positron emission tomography in patients with locally advanced non-small-cell lung cancer undergoing definitive chemoradiation therapy: Results of the ACRIN 6668/RTOG 0235 trial, J. Clin. Oncol. **31** (2013) 3823–3830.

[15] LU, Y.Y., et al., 18F-FDG PET or PET/CT for detecting extensive disease in small-cell lung cancer: A systematic review and meta-analysis, Nucl. Med. Commun. **35** (2014) 697–703.

[16] NATIONAL COMPREHENSIVE CANCER NETWORK, NCCN Guidelines: Small Cell Lung Cancer, NCCN, Plymouth Meeting, PA (2018).

[17] MITCHELL, M.D., AGGARWAL, C., TSOU, A.Y., TORIGIAN, D.A., TREADWELL, J.R., Imaging for the pretreatment staging of small cell lung cancer: A systematic review, Acad. Radiol. **23** (2016) 1047–1056.

[18] NATIONAL COMPREHENSIVE CANCER NETWORK, NCCN Guidelines: Malignant Pleural Mesothelioma, NCCN, Plymouth Meeting, PA (2018).

[19] KANEMURA, S., et al., Metabolic response assessment with $^{18}$F-FDG–PET/CT is superior to modified RECIST for the evaluation of response to platinum-based doublet chemotherapy in malignant pleural mesothelioma, Eur. J. Radiol. **86** (2017) 92–98.

[20] GERBAUDO, V.H., MAMEDE, M., TROTMAN-DICKENSON, B., HATABU, H., SUGARBAKER, D.J., FDG PET/CT patterns of treatment failure of malignant pleural mesothelioma: Relationship to histologic type, treatment algorithm, and survival, Eur. J. Nucl. Med. Mol. Imaging **38** (2011) 810–821.

[21] ELLIOTT, H.S., et al., $^{18}$F-FDG PET/CT in the management of patients with malignant pleural mesothelioma being considered for multimodality therapy: Experience of a tertiary referral center, Br. J. Radiol. **91** (2018) 20170814.

[22] KRAMMER, J., et al., (18)F-FDG PET/CT for initial staging in breast cancer patients — Is there a relevant impact on treatment planning compared to conventional staging modalities? Eur. Radiol. **25** (2015) 2460–2469.

[23] SUN, Z., YI, Y.L., LIU, Y., XIONG, J.P., HE, C.Z., Comparison of whole-body PET/PET–CT and conventional imaging procedures for distant metastasis staging in patients with breast cancer: A meta-analysis, Eur. J. Gynaecol. Oncol. **36** (2015) 672–676.

[24] PIVA, R., et al., Comparative diagnostic accuracy of $^{18}$F-FDG PET/CT for breast cancer recurrence, Breast Cancer **9** (2017) 461–471.

[25] TIAN, F., SHEN, G., DENG, Y., DIAO, W., JIA, Z., The accuracy of 18F-FDG PET/CT in predicting the pathological response to neoadjuvant chemotherapy in patients with breast cancer: A meta-analysis and systematic review, Eur. Radiol. **27** (2017) 4786–4796.

[26] HAYES, T., SMYTH, E., RIDDELL, A., ALLUM, W., Staging in esophageal and gastric cancers, Hematol. Oncol. Clin. North Am. **31** (2017) 427–440.

[27] LORDICK, F., et al., PET to assess early metabolic response and to guide treatment of adenocarcinoma of the oesophagogastric junction: The MUNICON phase II trial, Lancet Oncol. **8** (2007) 797–805.

[28] MONJAZEB, A.M., et al., Outcomes of patients with esophageal cancer staged with [$^{18}$F]fluorodeoxyglucose positron emission tomography (FDG–PET): Can postchemoradiotherapy FDG-PET predict the utility of resection? J. Clin. Oncol. **28** (2010) 4714–4721.

[29]  DONSWIJK, M.L., HESS, S., MULDERS, T., LAM, M.G., [18F]fluorodeoxyglucose PET/computed tomography in gastrointestinal malignancies, PET Clin. **9** (2014) 421–441, v–vi.

[30]  LU, Y.-Y., et al., Use of FDG–PET or PET/CT to detect recurrent colorectal cancer in patients with elevated CEA: A systematic review and meta-analysis, Int. J. Colorectal Dis. **28** (2013) 1039–1047.

[31]  MARCUS, C., SUBRAMANIAM, R.M., PET/computed tomography and precision medicine: Gastric cancer, PET Clin. **12** (2017) 437–447.

[32]  POULOU, L.S., et al., FDG–PET for detecting local tumor recurrence of ablated liver metastases: A diagnostic meta-analysis, Biomarkers **17** (2012) 532–538.

[33]  VAN KESSEL, C.S., et al., Preoperative imaging of colorectal liver metastases after neoadjuvant chemotherapy: A meta-analysis, Ann. Surg. Oncol. **19** (2012) 2805–2813.

[34]  METSER, U., et al., Effect of chemotherapy on the impact of FDG–PET/CT in selection of patients for surgical resection of colorectal liver metastases: Single center analysis of PET-CAM randomized trial, Ann. Nucl. Med. **31** (2017) 153–162.

[35]  CHEN, W., ZHUANG, H., CHENG, G., TORIGIAN, D.A., ALAVI, A., Comparison of FDG–PET, MRI and CT for post radiofrequency ablation evaluation of hepatic tumors, Ann. Nucl. Med. **27** (2013) 58–64.

[36]  MAAS, M., et al., What is the most accurate whole-body imaging modality for assessment of local and distant recurrent disease in colorectal cancer? A meta-analysis: Imaging for recurrent colorectal cancer, Eur. J. Nucl. Med. Mol. Imaging **38** (2011) 1560–1571.

[37]  XIA, Q., et al., Prognostic significance of (18)FDG PET/CT in colorectal cancer patients with liver metastases: A meta-analysis, Cancer Imaging **15** (2015) 19.

[38]  DOS ANJOS, D.A., et al., (18)F-FDG uptake by rectal cancer is similar in mucinous and nonmucinous histological subtypes, Ann. Nucl. Med. **30** (2016) 513–517.

[39]  MAHMUD, A., POON, R., JONKER, D., PET imaging in anal canal cancer: A systematic review and meta-analysis, Br. J. Radiol. **90** (2017) 20170370.

[40]  JONES, M., HRUBY, G., SOLOMON, M., RUTHERFORD, N., MARTIN, J., The role of FDG–PET in the initial staging and response assessment of anal cancer: A systematic review and meta-analysis, Ann. Surg. Oncol. **22** (2015) 3574–3581.

[41]  ASAGI, A., et al., Utility of contrast-enhanced FDG–PET/CT in the clinical management of pancreatic cancer: Impact on diagnosis, staging, evaluation of treatment response, and detection of recurrence, Pancreas **42** (2013) 11–19.

[42]  WILSON, J.M., MUKHERJEE, S., BRUNNER, T.B., PARTRIDGE, M., HAWKINS, M.A., Correlation of 18F-fluorodeoxyglucose positron emission tomography parameters with patterns of disease progression in locally advanced pancreatic cancer after definitive chemoradiotherapy, Clin. Oncol. **29** (2017) 370–377.

[43]  SCHICK, V., et al., Diagnostic impact of 18F-FDG PET-CT evaluating solid pancreatic lesions versus endosonography, endoscopic retrograde cholangio-pancreatography with intraductal ultrasonography and abdominal ultrasound, Eur. J. Nucl. Med. Mol. Imaging **35** (2008) 1775–1785.

[44]  KORN, R.L., et al., 18F-FDG PET/CT response in a phase 1/2 trial of nab-paclitaxel plus gemcitabine for advanced pancreatic cancer, Cancer Imaging **17** (2017) 23.

[45] HO, C.-L., YU, S.C., YEUNG, D.W., [11]C-acetate PET imaging in hepatocellular carcinoma and other liver masses, J. Nucl. Med. **44** (2003) 213–221.

[46] CHEUNG, T.T., et al., [11]C-acetate and [18]F-FDG PET/CT for clinical staging and selection of patients with hepatocellular carcinoma for liver transplantation on the basis of Milan criteria: Surgeon's perspective, J. Nucl. Med. **54** (2013) 192–200.

[47] CORVERA, C.U., et al., [18]F-fluorodeoxyglucose positron emission tomography influences management decisions in patients with biliary cancer, J. Am. Coll. Surg. **206** (2008) 57–65.

[48] NATIONAL COMPREHENSIVE CANCER NETWORK, NCCN Guidelines: Hepatobiliary Cancers, National Comprehensive Cancer Network, NCCN, Plymouth Meeting, PA (2018)

[49] SUN, D.-W., et al., Prognostic significance of parameters from pretreatment (18)F-FDG PET in hepatocellular carcinoma: A meta-analysis, Abdom. Radiol. **41** (2016) 33–41.

[50] LI, J., et al., Preoperative assessment of hilar cholangiocarcinoma by dual-modality PET/CT, J. Surg. Oncol. **98** (2008) 438–443.

[51] KIM, J.Y., et al., Clinical role of [18]F-FDG PET–CT in suspected and potentially operable cholangiocarcinoma: A prospective study compared with conventional imaging, Am. J. Gastroenterol. **103** (2008) 1145–1151.

[52] ESCUDIER, B., et al., Renal cell carcinoma: ESMO Clinical Practice Guidelines for diagnosis, treatment and follow-up, Ann. Oncol. **30** (2019) 706–720.

[53] MOTZER, R.J., et al., Kidney cancer: Clinical practice guidelines in oncology, J. NCCN **7** (2009) 618–630.

[54] ÖZTÜRK, H., Detecting metastatic bladder cancer using (18)F-fluorodeoxyglucose positron-emission tomography/computed tomography, Cancer Res. Treat. **47** (2015) 834–843.

[55] BECHERER, A., et al., FDG PET is superior to CT in the prediction of viable tumour in post-chemotherapy seminoma residuals, Eur. J. Radiol. **54** (2005) 284–288.

[56] VAN DE PUTTE, E.E.F., et al., FDG–PET/CT for response evaluation of invasive bladder cancer following neoadjuvant chemotherapy, Int. Urol. Nephrol. **49** (2017) 1585–1591.

[57] ARSENAULT, F., BEAUREGARD, J.-M., POULIOT, F., Prostate-specific membrane antigen for prostate cancer theranostics: From imaging to targeted therapy, Curr. Opin. Support Palliat. Care **12** (2018) 359–365.

[58] TREGLIA, G., et al., Diagnostic performance of fluorine-18-fluorodeoxyglucose positron emission tomography in the postchemotherapy management of patients with seminoma: Systematic review and meta-analysis, Biomed. Res. Int. **2014** (2014) 852681.

[59] SANDA, M.G., et al., Clinically localized prostate cancer: AUA/ASTRO/SUO guideline, Part II: Recommended approaches and details of specific care options, J. Urol. **199** (2018) 990–997.

[60] FENDLER, W.P., et al., [68]Ga-PSMA PET/CT: Joint EANM and SNMMI procedure guideline for prostate cancer imaging: Version 1.0, Eur. J. Nucl. Med. Mol. Imaging **44** (2017) 1014–1024.

[61] MIYAHIRA, A.K., et al., Meeting report from the Prostate Cancer Foundation PSMA-directed radionuclide scientific working group, Prostate **78** (2018) 775–789.

[62]   FANTI, S., et al., Development of standardized image interpretation for [68]Ga-PSMA PET/CT to detect prostate cancer recurrent lesions, Eur. J. Nucl. Med. Mol. Imaging **44** (2017) 1622–1635.

[63]   KHIEWVAN, B., et al., An update on the role of PET/CT and PET/MRI in ovarian cancer, Eur. J. Nucl. Med. Mol. Imaging **44** (2017) 1079–1091.

[64]   BJURLIN, M.A., ROSENKRANTZ, A.B., BELTRAN, L.S., RAAD, R.A., TANEJA, S.S., Imaging and evaluation of patients with high-risk prostate cancer, Nat. Rev. Urol. **12** (2015) 617–628.

[65]   SCHWARZ, J.K., GRIGSBY, P.W., DEHDASHTI, F., DELBEKE, D., The role of [18]F-FDG PET in assessing therapy response in cancer of the cervix and ovaries, J. Nucl. Med. **50** (2009) 1:64S–73S

[66]   COLOMBO, N., et al., ESMO-ESGO-ESTRO Consensus Conference on Endometrial Cancer: Diagnosis, treatment and follow-up, Ann. Oncol. **27** (2016) 16–41.

[67]   MUSTO, A., et al., Role of [18]F-FDG PET/CT in the carcinoma of the uterus: A review of literature, Yonsei Med. J. **55** (2014) 1467–1472.

[68]   GRIGSBY, P.W., PET/CT imaging to guide cervical cancer therapy, Future Oncol. **5** (2009) 953–958.

[69]   KHAN, S.R., ROCKALL, A.G., BARWICK, T.D., Molecular imaging in cervical cancer, Q. J. Nucl. Med. Mol. Imaging **60** (2016) 77–92.

[70]   HERRERA, F.G., PRIOR, J.O., The role of PET/CT in cervical cancer, Front. Oncol. **3** (2013) 34.

[71]   ROBERTSON, N.L., et al., The impact of FDG–PET/CT in the management of patients with vulvar and vaginal cancer, Gynecol. Oncol. **140** (2016) 420–424.

[72]   KASSEM, T.W., ABDELAZIZ, O., EMAD-ELDIN, S., Diagnostic value of 18F-FDG–PET/CT for the follow-up and restaging of soft tissue sarcomas in adults, Diagn. Interv. Imaging **98** (2017) 693–698.

[73]   SUBHAWONG, T.K., WINN, A., SHEMESH, S.S., PRETELL-MAZZINI, J., F-18 FDG PET differentiation of benign from malignant chondroid neoplasms: A systematic review of the literature, Skelet. Radiol. **46** (2017) 1233–1239.

[74]   NEMETH, Z., BOÉR, K., BORBÉLY, K., Advantages of [18]F FDG–PET/CT over conventional staging for sarcoma patients, Pathol. Oncol. Res. **25** (2017) 131–136.

[75]   CHESON, B.D., PET/CT in lymphoma: Current overview and future directions, Semin. Nucl. Med. **48** (2018) 76–81.

[76]   SCHÜLE, S.-C., et al., Influence of (18)F-FDG PET/CT on therapy management in patients with stage III/IV malignant melanoma, Eur. J. Nucl. Med. Mol. Imaging **43** (2016) 482–488.

[77]   NATIONAL COMPREHENSIVE CANCER NETWORK, NCCN Guidelines: Melanoma, Version 2, NCCN, Plymouth Meeting, PA (2018)

[78]   MENA, E., SANLI, Y., MARCUS, C., SUBRAMANIAM, R.M., Precision medicine and PET/computed tomography in melanoma, PET Clin. **12** (2017) 449–458.

[79]   SCHRÖER-GUNTHER, M.A., et al., F-18-fluoro-2-deoxyglucose positron emission tomography (PET) and PET/computed tomography imaging in primary staging of patients with malignant melanoma: A systematic review, Syst. Rev. **1** (2012) 62.

[80]  CISTARO, A., et al., Italian multicenter study on accuracy of [18]F-FDG PET/CT in assessing bone marrow involvement in pediatric Hodgkin lymphoma, Clin. Lymphoma Myeloma Leuk. **18** (2018) e267–e273.

[81]  BARRINGTON, S.F., et al., Role of imaging in the staging and response assessment of lymphoma: Consensus of the International Conference on Malignant Lymphomas Imaging Working Group, J. Clin. Oncol. **32** (2014) 3048–3058.

[82]  SPECHT, L., BERTHELSEN, A.K., PET/CT in radiation therapy planning, Semin. Nucl. Med. **48** (2018) 67–75.

[83]  EL-GALALY, T.C., GORMSEN, L.C., HUTCHINGS, M., PET/CT for staging: Past, present, and future, Semin. Nucl. Med. **48** (2018) 4–16.

[84]  CAVO, M., et al., Role of [18]F-FDG PET/CT in the diagnosis and management of multiple myeloma and other plasma cell disorders: A consensus statement by the International Myeloma Working Group, Lancet Oncol. **18** (2017) e206–e217.

[85]  MOON, S.H., et al., Prognostic value of baseline [18]F-fluorodeoxyglucose PET/CT in patients with multiple myeloma: A multicenter cohort study, Korean J. Radiol. **19** (2018) 481–488.

[86]  STOLZENBURG, A., et al., Prognostic value of [[18]F] FDG–PET/CT in multiple myeloma patients before and after allogeneic hematopoietic cell transplantation, Eur. J. Nucl. Med. Mol. Imaging **45** (2018) 1694–1704.

[87]  KIM, S.-J., LEE, S.-W., PAK, K., SHIM, S.-R., Diagnostic performance of PET in thyroid cancer with elevated anti-Tg Ab, Endocr. Relat. Cancer **25** (2018) 643–652.

[88]  CENGIZ, A., et al., Correlation between baseline [18]F-FDG PET/CT findings and CD38- and CD138-expressing myeloma cells in bone marrow and clinical parameters in patients with multiple myeloma, Turk. J. Haematol. **35** (2018) 175–180.

[89]  SANTHANAM, P., KHTHIR, R., SOLNES, L.B., LADENSON, P.W., The relationship of Braf(V600e) mutation status to FDG PET/CT avidity in thyroid cancer: A review and meta-analysis, Endocr. Pract. **24** (2018) 21–26.

[90]  TAKEUCHI, S., et al., Impact of 18 F-FDG PET/CT on the management of adrenocortical carcinoma: Analysis of 106 patients, Eur. J. Nucl. Med. Mol. Imaging **41** (2014) 2066–2073.

[91]  SCHÜTZ, F., LAUTENSCHLÄGER, C., LORENZ, K., HAERTING, J., Positron emission tomography (PET) and PET/CT in thyroid cancer: A systematic review and meta-analysis, Eur. Thyroid J. **7** (2018) 13–20.

[92]  PATTISON, D.A., et al., [18]F-FDG-avid thyroid incidentalomas: The importance of contextual interpretation, J. Nucl. Med. **59** (2018) 749–755.

[93]  YEH, M.W., et al., American Thyroid Association statement on preoperative imaging for thyroid cancer surgery, Thyroid **25** (2015) 3–14.

[94]  SMALLRIDGE, R.C., DIEHL, N., BERNET, V., Practice trends in patients with persistent detectable thyroglobulin and negative diagnostic radioiodine whole body scans: A survey of American Thyroid Association members, Thyroid **24** (2014) 1501–1507.

[95]  BOZKURT, M.F., et al., Guideline for PET/CT imaging of neuroendocrine neoplasms with [68]Ga-DOTA-conjugated somatostatin receptor targeting peptides and [18]F-DOPA, Eur. J. Nucl. Med. Mol. Imaging **44** (2017) 1588–1601.

[96]   HÖRSCH, D., et al., Neuroendocrine tumors of the bronchopulmonary system (typical and atypical carcinoid tumors): Current strategies in diagnosis and treatment. Conclusions of an expert meeting February 2011 in Weimar, Germany, Oncol. Res. Treat. **37** (2014) 266–276.

[97]   SUNDIN, A., et al., ENETS consensus guidelines for the standards of care in neuroendocrine tumors: Radiological, nuclear medicine and hybrid imaging, Neuroendocrinology **105** (2017) 212–244.

[98]   BAUMANN, T., ROTTENBURGER, C., NICOLAS, G., WILD, D., Gastroenteropancreatic neuroendocrine tumours (GEP–NET) — Imaging and staging, Best Pract. Res. Clin. Endocrinol. Metab. **30** (2016) 45–57.

[99]   BAR-SEVER, Z., et al., Guidelines on nuclear medicine imaging in neuroblastoma, Eur. J. Nucl. Med. Mol. Imaging **45** (2018) 2009–2024.

[100]  PANDIT-TASKAR, N., MODAK, S., Norepinephrine transporter as a target for imaging and therapy, J. Nucl. Med. **58** Suppl 2 (2017) 39S–53S.

[101]  KAN, Y., et al., [68]Ga-somatostatin receptor analogs and [18]F-FDG PET/CT in the localization of metastatic pheochromocytomas and paragangliomas with germline mutations: A meta-analysis, Acta Radiol. **59** (2018) 1466–1474.

[102]  TAIEB, D., et al., EANM 2012 guidelines for radionuclide imaging of phaeochromocytoma and paraganglioma, Eur. J. Nucl. Med. Mol. Imaging **39** (2012) 1977–1995.

[103]  TREGLIA, G., TAMBURELLO, A., GIOVANELLA, L., Detection rate of somatostatin receptor PET in patients with recurrent medullary thyroid carcinoma: A systematic review and a meta-analysis, Hormones **16** (2017) 362–372.

[104]  WELLS, S.A., Jr., et al., Revised American Thyroid Association guidelines for the management of medullary thyroid carcinoma, Thyroid **25** (2015) 567–610.

[105]  SANTHANAM, P., et al., Nuclear imaging of neuroendocrine tumors with unknown primary: Why, when and how? Eur. J. Nucl. Med. Mol. Imaging **42** (2015) 1144–1155.

[106]  PARK, S.B., et al., Role of [18]F-FDG PET/CT in patients without known primary malignancy with skeletal lesions suspicious for cancer metastasis, PLoS One **13** (2018) e0196808.

[107]  TAMAM, C., TAMAM, M., MULAZIMOGLU, M., The accuracy of [18]F-fluorodeoxyglucose positron emission tomography/computed tomography in the evaluation of bone lesions of undetermined origin, World J. Nucl. Med. **15** (2016) 124–129.

# ABBREVIATIONS

| | |
|---|---|
| CT | computed tomography |
| DOPA | dihydroxyphenylalanine |
| DOTA | dodecane tetra-acetic acid |
| FDG | fluorodeoxyglucose |
| FDOPA | fluoro-l-dihydroxyphenylalanine |
| FES | fluoroestradiol |
| FET | fluoroethyltyrosine |
| GEP-NET | gastroenteropancreatic neuroendocrine tumours |
| MRI | magnetic resonance imaging |
| NET | neuroendocrine tumour |
| NSCLC | non-small cell lung carcinoma |
| PET | positron emission tomography |
| PRRT | peptide receptor radionuclide therapy |
| PSMA | prostate specific membrane antigen |
| SCLC | small cell lung carcinoma |
| SSA | somatostatin analogue |

# CONTRIBUTORS TO DRAFTING AND REVIEW

| | |
|---|---|
| Dondi, M. | University of Brescia, Italy |
| Doruyter, A. | Tygerberg Hospital, South Africa |
| Estrada Lobato, E. | International Atomic Energy Agency |
| Giammarile, F. | International Atomic Energy Agency |
| Lee, S.T. | Austin Health, Australia |
| Macapinlac, H. | M.D. Anderson Cancer Center, United States of America |
| Mariani, G. | University of Pisa, Italy |
| Navarro-Marulanda, M.C. | International Atomic Energy Agency |
| Paez, D. | International Atomic Energy Agency |
| Rangarajan, V. | Tata Memorial Hospital, India |
| Rodriguez Sanchez, Y. | International Atomic Energy Agency |

# ORDERING LOCALLY

IAEA priced publications may be purchased from the sources listed below or from major local booksellers.

Orders for unpriced publications should be made directly to the IAEA. The contact details are given at the end of this list.

## NORTH AMERICA

*Bernan / Rowman & Littlefield*
15250 NBN Way, Blue Ridge Summit, PA 17214, USA
Telephone: +1 800 462 6420 • Fax: +1 800 338 4550
Email: orders@rowman.com • Web site: www.rowman.com/bernan

## REST OF WORLD

Please contact your preferred local supplier, or our lead distributor:

*Eurospan Group*
Gray's Inn House
127 Clerkenwell Road
London EC1R 5DB
United Kingdom

*Trade orders and enquiries:*
Telephone: +44 (0)176 760 4972 • Fax: +44 (0)176 760 1640
Email: eurospan@turpin-distribution.com

*Individual orders:*
www.eurospanbookstore.com/iaea

*For further information:*
Telephone: +44 (0)207 240 0856 • Fax: +44 (0)207 379 0609
Email: info@eurospangroup.com • Web site: www.eurospangroup.com

**Orders for both priced and unpriced publications may be addressed directly to:**
Marketing and Sales Unit
International Atomic Energy Agency
Vienna International Centre, PO Box 100, 1400 Vienna, Austria
Telephone: +43 1 2600 22529 or 22530 • Fax: +43 1 26007 22529
Email: sales.publications@iaea.org • Web site: www.iaea.org/publications